SpringerBriefs in Geography

SpringerBriefs in Geography presents concise summaries of cutting-edge research and practical applications across the fields of physical, environmental and human geography. It publishes compact refereed monographs under the editorial supervision of an international advisory board with the aim to publish 8 to 12 weeks after acceptance. Volumes are compact, 50 to 125 pages, with a clear focus. The series covers a range of content from professional to academic such as: timely reports of state-of-the art analytical techniques, bridges between new research results, snapshots of hot and/or emerging topics, elaborated thesis, literature reviews, and in-depth case studies.

The scope of the series spans the entire field of geography, with a view to significantly advance research. The character of the series is international and multidisciplinary and will include research areas such as: GIS/cartography, remote sensing, geographical education, geospatial analysis, techniques and modeling, landscape/regional and urban planning, economic geography, housing and the built environment, and quantitative geography. Volumes in this series may analyze past, present and/or future trends, as well as their determinants and consequences. Both solicited and unsolicited manuscripts are considered for publication in this series. SpringerBriefs in Geography will be of interest to a wide range of individuals with interests in physical, environmental and human geography as well as for researchers from allied disciplines.

Girish Chandrakant Gujar • Adolf K. Y. Ng

Blue Economy and Smart Sea Transport Systems

Maritime Security

Girish Chandrakant Gujar
Vijay Patil School of Management
D Y Patil University
Mumbai, India

Adolf K. Y. Ng
Faculty of Business and Management
BNU-HKBU United International College
Zhuhai, Guangdong Province, China

ISSN 2211-4165   ISSN 2211-4173   (electronic)
SpringerBriefs in Geography
ISBN 978-3-031-21633-6   ISBN 978-3-031-21634-3   (eBook)
https://doi.org/10.1007/978-3-031-21634-3

This Springer imprint is published by the registered company Springer Nature Switzerland AG
The registered company address is: Gewerbestrasse 11, 6330 Cham, Switzerland

Preface

There is no gainsaying by restating or emphasizing the fact that the world is heading relentlessly with increasing speed towards unalterable global societal changes. However, it remains to be seen whether the destruction is creative or not. The changes are primarily driven by increased but selective globalization, climate change, and technological disruption. It is obvious that the global change will be non-strategic, unsustainable, as well as unequitable. All these changes are causing disgruntlement, resentment, and rage among the populace all over the world. They demand to be redressed immediately as they lead to visible collapse of societies, such as Venezuela, Zimbabwe, Afghanistan, Syria, Iraq, Yemen, and elsewhere, and also in emerging economies like India where the collapse is not yet manifestly evident. The list is just too long to be mentioned here. One gets to see elements of such collapse even in so-called developed countries too with identical reaction from the populace.

Asia, and particularly the Indian Ocean rim, is where the final act of the play is going to be staged. This is where the interplay between China, India, and the US, the top three nations (economic, population, and otherwise), happens and has far-reaching and long-lasting consequences globally. Whether it will be beneficial or not, only time will tell. However, being unprepared for the consequences will be, to say the least, foolish. In this sense, the book attempts to warn, prevent, and potentially avoid such foolishness.

Currently, the Indian Ocean is viewed as one of the most "active" oceans in the world, thanks to the extensive use of its waters for trade and transportation reaching out to the entire world. It hosts some of the world's most important (but also highly vulnerable) chokepoints and narrow passages, such as the Straits of Malacca, Hormuz, and Mandeb.

To address such, we need to develop a new, "Smart" global system where efficiency, adaptability, improvisation, innovation, and customer satisfaction matter more than size, growth rate, or revenue generation. It needs to possess the ability to consistently identify innovative ways to cut costs that improve security and eliminate waste. To its end, value addition is a mission rather than just a process. This heavily depends on substantially improved process efficiencies which, in turn, would lead to the evolved state of the *blue economy* underpinned by the concepts of re-cycling, waste reduction, and reduced pollution. To achieve such, we need to look beyond *narrow national* maritime *boundaries* towards the security of the entire global maritime commons. With maritime disputes on the rise, the world can no longer take security and the openness of the maritime commons for granted. We are moving towards very interesting time indeed.

Mumbai, India   Girish Chandrakant Gujar
Zhuhai, Guangdong Province, China   Adolf K. Y. Ng
January 2023

Contents

Chapter 1
Introduction

1.1 Setting the Scene

With a focus on maritime security, this book discusses the Blue Economy and the Smart sea transport systems in the context of the Indian Ocean rim countries and the multifaceted challenges they are likely to face while attempting to develop such systems. Two novel ideas emerged in during the 1950s in the US. The first idea of "globalization" was aggressively promoted, while the second idea of "limits to growth" was largely glossed over. It was only recently that the latter got its due in the face of the financial tsunami that took place in 2008. Currently, it appears some countries are facing the consequences of having overshot their limits to growth by way of global warming, while the rest have just not been able to achieve their potential and are facing rising unemployment and inequality in their societies (Reinhart & Rogoff, 2010).

Thus, we perforce witness mutation of sectors, regions, and countries to a higher evolved state due to the playacting of the above-named forces and those entities, which do not evolve will shrivel and die. This evolved situation will influence the formation of the "Blue Economy." It emphasizes lower costs, reduced waste, greater efficiency, and value addition in all value addition services particularly those activities associated with the near and deep oceans in all its manifestations and avatars. Also, it implies lesser production, reduced consumption, and near sourcing. Most importantly, it signifies greater security due to better manageability. Simultaneously, the sea transport systems involved in global trade too had to upgrade with similar objectives, thus the struggle for achieving a "Smart' sea transport system.

1.2   Background and Framework

Since the 1960s, most of the globally traded goods have been moved by container shipping. The supply chains have become more global as more and more operations are outsourced and moved offshore.[1] The impact of transport on supply chain performance has also increased. Thus, container transport has become the lifeline of almost all the global supply chains. In this case, a variety of issues concerning maritime transport have grabbed increasing attention (Woxenius et al., 2004). These include maritime security, greater vulnerability, potential delays, increased information provisioning, and greater congestion. Moreover, it leads to increased duration and variability of travel time. Since the 1990s, the world's container traffic has grown substantially in terms of gross domestic product (GDP). Nowadays, the world's annual container throughput is close to 800 million TEUs (UNCTAD, 2019/2021).

Terminal operators play major roles in handling and transporting containers by different transport modes mostly under the trunk-and-feeder system. Terminals are mostly engaged by shipping companies rather than by the shippers or consignees. On the other hand, the terminal operator has no contractual dealings with the inland transporters of containers. They are instead appointed by the shipping companies or the shippers/consignees. As such, it becomes challenging for the terminal operators to deal with the inland transport contractors. This is because a terminal operator must coordinate with the inland transporters and operational coordination of this kind becomes a challenging task as it involves multiple decision-makers. Thus, the success of global supply chains largely depends upon the smooth and seamless coordination between the different stakeholders in addition to higher efficiency, better service quality, and cost reduction asked for by ever more demanding customers (Dobbs & Manyika, 2015). The Smart Sea Transport System will be highly integrated, multirole, multinational entities. Such entities would interact with similar entities in different regions, which would result in creation of synergies for the benefits of their customers. The dynamics between the Blue Economy, maritime security, and the Smart Sea Transport System can be found in Fig. 1.1.

The Blue Economy is driving changes to the structure, distribution, and logics used in the Smart Sea Transport System. This change is moderated by the changing the structure, nodes of power and logics, used in the maritime security, implying change at a level above a single organization. In this case, two levels of analysis come to mind: population and institutional system. Hence, we look at the issue from the institutional perspective: a changing theoretical area that explores how institutional entrepreneurs refocus attention, change logics, alter institutional structure, and revise institutional order (cf. Thornton et al., 2012). Based on this framework,

---

[1] In addition to offshoring, in the past few years, the world has witnessed the gradual increase of reshoring (the return of productions/supply chains to home countries) and "friend"-shoring (the relocation of productions/supply chains to countries/regions that are deemed "reliable" or "trustworthy").

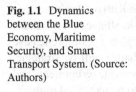

Fig. 1.1 Dynamics between the Blue Economy, Maritime Security, and Smart Transport System. (Source: Authors)

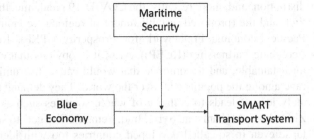

this book appeals to a wide audience in management, governance, and organizational behaviors.

This book answers the three "how" above from a systemic perspective, rather than from a ground-level one alone. This is a relatively underexplored area given the relative newness of the Blue Economy book from 2004, the World Bank report on the Potential of the Blue Economy in 2017, and the increasingly complex maritime security regulatory context including the Maritime Transportation Security Act of 2002 in the US. In this way, we examine the ways in which stakeholders can work altogether to build a more secure, more resilient global system of maritime trade. Hence, this book clarifies the effect of the emerging Blue Economy on the transport system and offers readers a greater understanding of the role maritime security plays in this changing relationship. Thus, people working in the maritime sector and people working in the security sector will have substantial interests in this book's contents as they possess some unique features:

1. Pioneering contributions by investigating the role of maritime security in the future Smart sea transport systems.
2. Systematic contributions to maritime security that focus on the dynamic relationships between maritime security and the Blue Economy.
3. Insightful conceptual analysis by focusing on individual contributions.
4. Real-world experience from different countries and regions, notably emerging markets, contributed by reputable scholars who are leaders in the research of this area.

The issue maritime security cannot be fully understood without taking reference to the triangular relationship between the US, China, and India. An overview of the recent relationship between these countries can be found below.

1.3 A Tango of Three: The Recent Triangular Relationships Between the US, China, and India

There is no gainsaying by restating or strongly emphasizing that the world is heading relentlessly toward unalterable global societal changes. The changes are primarily driven by increased but selective globalization, climate change, technological disruption, and, more recently, the COVID-19 pandemic, the Russo-Ukrainian conflict, and the (proposed) establishment of regional economic alliances (e.g., Indo-Pacific Economic Framework for Prosperity (IPEF), Regional Comprehensive Economic Partnership (RCEP)). Also, it is obvious that change will be strategic, unsustainable, and unequitable that would cause disgruntlement, resentment, and rage among the populace all over the world. They demand to be redressed immediately least it leads to collapse of whole societies such as what we witness (e.g., Zimbabwe, Afghanistan, Syria, Iraq, Yemen). One gets to see elements of such collapse even in so-called developed countries too with identical reaction from the populace.

Being the world's most populous continent, Asia is where the final act of the play will be staged, particularly in the Indian Ocean rim. This is where the interplay between the top three countries (economy and population and otherwise) will happen, which will pose consequences to the world for a long time. The recent border conflict between China and India in June 2020 serves as an illustrative example. Whether they will be beneficial or not only time will tell, but to leave it unprepared for the consequences would be injudicious.

In the US, China, India and elsewhere, a major policymaking problem is the choice between keeping "bankrupt" industries and sectors alive to enable excessive and redundant labors to survive or to give them time to retrain and shift to the more profitable new industries (e.g., ICT, renewable energy). The continuous existence of such excessive and redundant labors serves as a key factor explaining why the economic growth is held back. The second reason is the starvation of the sunrise industry of critically valuable capital being consumed by the archaic old industries, which further drag the economic growth southward. This lack of adequate economic growth leads to low availability of suitable employment, giving rise to host of issues (e.g., immense migration, pollution, poverty, disease, nationalism, demagoguery, intolerance). All these result in disgruntlement, resentment, and rage among populace, which occasionally erupts wherever and whenever the circumstances permit.

However, despite common wisdom, the US under the Trump administration seems to revert to its fundamental temperament—becomes less liberal and market oriented and tries to focus on domestic politics and economic growth. The potential trade war between US and China is claimed to "bring production back to North America." This will naturally impact the currently established global supply chain system—where the manufacturing of finished/semifinished products is often outsourced to emerging economies/developing countries. Moreover, the COVID-19 pandemic has led to some calls shifting the established "just-in-time" to a more "just-in-case" production system (Ng & Liu, 2014). However, given the higher labor costs involved in moving production of (semi-)finished products to North America, this will, inevitably, enhance the production costs of even the most basic commodities and hurt the common man in the pocket where it hurts the most. As such, the disruption of global trade benefits none even notionally.

Furthermore, with research and development (R&D) being one of US' major advantages, we expect that there will be more attempts among US manufacturers to develop a supply chain and production system based on nonlabor (or, at least, minimum labors) means. The major intention of the US to initiate a trade war with China is to restrict the latter from developing high-tech industries, say, artificial intelligence (AI), space, genetics, and robotics. The logic is simple: by keeping China from developing high-tech industries, China would find itself vulnerable to a deteriorating trade relationship with the US, as it is difficult to develop alternative production and supply chain systems in a short period of time. Indeed, many current US trade policies are already designed to achieve such an objective. Of course, this is a problem that demands immediate responses by China, as higher taxation and the lack of alternatives imply a wastage of money and loss of competitive advantage. Ultimately, it will be the end-users bearing the costs, with higher expenses on even the most basic commodities. A preference for long-term pains in exchange for short-term gains.

Domestically, we need to ascertain if we can achieve equitable economic growth in our society and what costs are involved. All the three countries, namely, the US, China, and India, face dilemmas as to whether they should ignore world politics and economics or attempt to influence global affairs and, if so, at what costs. In this regard, temperamentally, all three of them are rather "isolationists" and not particularly keen to engage with the world (unlike, for instance, many of the EU member states), thus creating a political vacuum yet to be filled. The results can be tragic as it can lead to a domino effect.

Moreover, since the rise (and fall) of Trump, the US is experiencing a retreat of the liberal ideology (Stephens, 2015). The policies put together by the Trump administration were essentially against the liberal and open global order that the country stood for in the past, which had strong views on social welfare. In addition, it was obviously aware of the huge population of China and India, and that both countries possess rich civilizations and cultural strength that can make them strong enough to tackle the US hegemony in the global political economy. Unsurprisingly, the US would like to restrain, if not eliminate, the rise of such a tripolar system, thus explaining the US' current foreign policy direction. It is like US relationship with Japan before World War II, where the former tried to limit the latter's involvement in global politics and economy. Similarly, it would now try to limit China's role in international politics and economy, especially the Middle East and East Asia, in exchange for trade concessions. This is exactly where the US and China, and probably India, are drawn into potentially serious conflicts.

While India does not have many choices due to the adopted democratic model of governance, China does have choices in facing such a challenge. Hence, China should ensure that they are not giving the US and its close allies the opportunity to come together against it (the latest agreement between the US, Canada, and Mexico—the USMCA with terms highly favorable to the Trump's administration served as an early warning, as well as the impacts posed by the COVID-19 pandemic on global supply chains, or more exactly, global value chains). In our view, China should focus on probable state of world after 2025. By doing so, it would play a significant role in promoting the liberal and free economic order of the past and the future economic and social well-being.

1.4   Structure of the Book

Chapter 2 introduces the concept of maritime security. It highlights why maritime security matters in the Blue Economy nowadays. Also, different aspects of maritime security and international maritime trade are discussed. The contents of this chapter include (1) the relevance of maritime security, (2) maritime security as a nationalistic political issue, (3) maritime terrorism, (4) maritime pollution and environmental disasters, and (5) container security. Today, it is an uncontestable and well-acknowledged fact that the oceans are where the future wealth is stored in the form of food, energy resources, and rare minerals. The oceans also support major sea lanes of communication and ports as trading hubs. History has also taught us that dominance of these sea lanes leads to obtaining global dominance and political power. Most of the sea lanes with their intermittent chokepoints pass through the Indian Ocean. Hence, this strategic location enjoys pre-eminence in the eyes of the major global powers. With almost 800 million containers being transported annually (UNCTAD, 2019/2021), the impact of transport on supply chain performance has increased tremendously. In recent times, several important issues connected with the maritime transportation of manufactured goods and raw materials have drawn attention of national governments all over the world. Also, one should remember that we now live in an era where customer satisfaction matters much more than growth rate, profit maximization, or revenue generation.

Chapter 3 investigates maritime piracy and related jurisdictional issues in today's global shipping, supply chains, and exclusive economic zones. Maritime piracy is a complex and an intricate problem that is often the manifestation of various socio-political problems that ail a particular geographical area. The contents of this chapter include (1) the importance of tackling piracy in the Blue Economy and Smart sea transport systems, (2) pirates as a socio-political problem, (3) piracy and container security, (4) tools in tackling piracy and their effectiveness, and (5) the way forward. Currently, the Indian Ocean is viewed as the most "active" ocean as compared to the Atlantic or Pacific. This is mainly because it is hosting several activities and players from all over the world ranging from extensive trade, transportation, which is much larger in scope and size due to availability of large markets/populations and cheap raw materials in abundance such as oil and natural gas. Significantly, some of the world's most important choke points and narrow passages such as the straits of Malacca, Hormuz, and Mandeb along with their associated vulnerabilities are also located in this region. However, despite the nonconformity and noncompatibility and divergence on many issues in the region, both India and China are willing to shoulder responsibilities of global policing of maritime commons, particularly in the Indian Ocean. These responsibilities were earlier discharged by the US. Considering the limitations and the maritime capacities of the other littorals, it has become imperative for the capable and willing nations such India and China to give a helping hand to the US to discharge their responsibilities. This clearly underlies the areas of agreements as well as disagreements and view the security of the global maritime commons from a holistic perspective. In addition, all the three powers are keen to project their sovereign political power beyond the physical limits as proscribed by the Convention on Law of the Sea (UNCLOS). The objective of the three powers is to secure their future international trade movement and exploit the riches of the deep oceans. The absence of adequate ratification and implementation of the UNCLOS convention could lead to a clash of the vested interests of the three powers. In such circumstances, real physical conflict among the three cannot be ruled out.

Chapter 4 investigates the major challenges for coastal security. Coastal security is extremely important as they serve as the "first line of defense" against terrorist and other illegal activities getting into any countries/regions. A thorough discussion will take place on how coastal security is addressed, notably the balance between maximum security versus strategic risk-taking. The contents include (1) the concept of coastal security and its importance to the Blue Economy and Smart sea transport systems, (2) early attempts at coastal security, (3) measures to upgrade coastal security, (4) infrastructural resources, and (5) cooperation between levels of governments and intergovernmental organizations. Also, it investigates security at ports and inland ports, including (1) the roles of ports and inland ports in providing maritime security, (2) the classification of ports and inland ports. It also discusses the specific liabilities of the ports and dry ports while discharging their duties. (3) The national regulatory competencies. (4) The obstacles and challenges of the ports and dry ports in securing maritime security. A glaring issue concerning hinterland security is related to that of supply chain security or what could be colloquially termed as container security. Today, container security getting compromised is a common occurrence mostly due to erroneous information being provided by the stakeholders either by accident or design, but none of the stakeholders are saddled with the legal responsibility of verifying the correctness of the information provided.

Chapter 5 investigates extraordinary technological disruption and its impact on maritime trade due to various reasons including the recently raging Covid-19 pandemic, which disrupts the supply and demand equation, causing extreme volatility in freight rates and immense delays and congestion. Establishing intelligent infrastructure is imperative to ensure smooth and efficient traffic flows and trade flows in the Smart system. An intermodal traffic center will interlink the various modes of transport and will also ensure the efficiency of traffic flows. Such a Centre will process all traffic information collected in the smart port in real time and distribute it to all stakeholders automatically.

Chapter 6 provides specific, real-world cases on how governments around the world tackle the challenges of maritime security and their roles in achieving the Blue Economy and the future Smart sea transport systems. Examples will be taken from major economic powerhouses and countries located in strategic locations. Today's problem is not economic growth alone but more equitable and sustainable economic growth. There is also the issue of jobless growth. How to train the populace in the necessary skills required for quality jobs in the face of automation compounds the challenge further. Also, who to fund this growth and to what extent the funds will be made available (and to whom) from the government is another question. The role of corporates (the ones who will probably provide employment) is yet to be decided. The question of climate change and sustainability in all its manifestations and myriad is also staring in the face floating in the air as well as water. Health epidemics caused by climate change, either directly or indirectly, is challenging the policymakers all over the world, as we are already witnessing since 2020.

Chapter 7 discusses the development of maritime clusters and their role in promotion of blue economy. Essentially maritime clusters are nothing but an endeavor by several nations searching for systematic synergies in various maritime activities, such as ship building, port development, promotion of maritime education, green shipping finance, and coastal tourism.

Chapter 8 illustrates the importance of port-land interface–related projects for the development of blue economy. It consists of an analysis of the current scenario— How and Why of Port-Land Interface Policy and the future scenario—How and why it should be altered. The chapter ends with some useful recommendations for further improvements.

Finally, Chap. 9 discusses the impacts of the Covid-19 pandemic on the blue economy and the way forward. It looks at the impact on international maritime trade, ports, and the Green Regulations. Towards the end, it explores the appropriate way in the development of the Blue Economy in the future.

Acknowledgments The support from CCAPPTIA (ccapptia.com) is gratefully acknowledged. Also, some parts of the contents in this book have been published previously by the authors in our previous publications. The details of our earlier publications are as follows:

1. Gujar, G. C., & Hong Yan. (2013). "Smart Port in Smart Era" Maritime Insight issue published by the C Y Tung International Centre of Maritime Studies of the Hong Kong Polytechnic University in September 2013.
2. Gujar, G. C., Ghosh, P. K., & Hong Yan. (2014). "Securing the Maritime Commons", Maritime Insight issue published by the C Y Tung International Centre of Maritime Studies of the Hong Kong Polytechnic University in December 2014.
3. "Contemporary Container Security" (ISBN 978-3-319-98133-8) by Girish Gujar, Adolf K.Y. Ng and Zaili Yang, Palgrave Macmillan, UK.

Chapter 2
Maritime Security: Call of the Seas

Abstract This chapter discusses the concept of maritime security. In today's world, maritime and economic security goes hand in hand and the term "security" has become ubiquitous for contemporary policymakers. The question that needs urgently answering is how it should be conceived and how should it be enacted and implemented. Furthermore, what are the policy implications? This chapter addresses these queries.

Keywords Maritime security · International trade · Policy implications · Institutions

2.1 The Concept of Security

The term "security" has become ubiquitous for contemporary policymakers. In both academic and policymaking circles, confusion is abound with regard to its definition, which signifies different things or conditions to different people in differing circumstances. The question hence that needs answering is how it should be conceived and how should it be enacted and implemented. Furthermore, what are the policy implications?

In *The Evolution of International Security Studies*, Barry Buzan views security at multidimensional levels of analysis such as individual, national, and international (both regional and system wide) security. Also, he views it from the military, political, societal, economic, and environmental perspectives.

The security of individuals and of the community is a sole function of the state as well as the supra state. As such, it is the responsibility of such entities to protect the individual and the group, which is nothing but group of individuals to externally and internally generated "threats." The state may discharge the security function by creating an environment that enhances the rule of law. Furthermore, the state would ensure equal economic opportunities for all as well as provide social welfare facilities. However, one cannot overlook the irony that more security that is sought by individuals (and provided by the state), the greater the extent to which freedom must be compromised and vice versa. This is particularly true for the forthcoming post-covid-19 era. In defining security as a measure of the absence of threats to specifically stated values by the government, Bichou (2004) indicates the potential for confusion when the symbol of national security is "used in a generic manner."

Finally, we should realize that eventually, everybody will look at behavior of states and the systemic factors, which influence their behavior while reviewing global security policies. As such, the definition rests on the fundamental assumption that the states hold primary responsibility for providing security to their respective populations though some international organizations may also play supporting roles in the provision of security to various communities at different levels.

2.2 Relevance of Maritime Security

In today's world, maritime and economic security goes hand in hand. According to an International Maritime Organization (IMO) report on climate change and livelihoods, in Southeast Asia alone, over 200 million people depend upon the sea for their livelihoods. In addition, several maritime industries also dominate the economies of their respective countries such as ship building, shipbreaking, offshore oil and gas, tourism, fishing, and trade.

Through international trade—mostly carried by sea—the process of trade is the most efficient when it is relatively free with states and regions demanding a higher degree of autonomy under the garb of "national interest." It should also be noted that even Adam Smith in his *Wealth of Nations* recognized the primary importance of maritime security as necessary for the defense of national assets. Hence, he advised the promulgation of international laws, such as the freedom of Navigation Acts, protection of fisheries, and preserving the monopolies of trade granted by the sovereign.

The major advantage in this aspect is the spreading awareness among the popular and a new maritime development strategy, adopted by several countries. This strategy aggressively promotes the maritime technology industry, along with communal exploitation of ocean thermal energy and mining of seabed minerals.

2.3 Maritime Security: A Nationalistic Political Issue

At present the north-western end of Indian Ocean region is currently perceived to be the center of strategic gravity of the world. This region is subject to extensive "proxy politics" playing host to several small and great powers. The erosion of US naval power in this region may be notional, but the jostle for power is real. The main reason for this conundrum is that major revisionist players, such as China, India, and Russia, are seeking to enhance their strategic influence by looking for primacy along with that of a reluctant and isolationist USA.

The current trend of de-globalization has placed seaborne trade in parallel with transnational threats. Escalating incidents of piracy, terrorism, drug trafficking, illegal immigration, and arms running in their ever-evolving manifestations have emerged as the bane of seafarers. The incidents of smuggling contraband regularly defeat the purpose of national economic and trade policies. These maritime security challenges are essentially asymmetric in nature and hence are difficult and costly to tackle. Another issue that complicates the situation further is the considerable dissimilarities between capabilities and competencies of the littorals.

Due to the traversing of nearly 100,000 ships carrying all kinds of cargoes annually, the ocean is very busy indeed. In addition, the straits of Malacca and Hormuz handle over 80% of these ships making the region even more important. All these heavily underline the importance of the need for an adequate and appropriate maritime security policy.

2.4 Piracy and Maritime Terrorism

The rise in seaborne trade has adversely impacted the security of sovereign states by all kinds of threats including asymmetric ones such as piracy. This also enhances the transactional costs of trading as well as influencing the economic indices of a trade-dependent state and its economic development.

To tackle such issues, the UNSC was forced to adopt *Resolution 1851(2009)* in January 2009. This resolution established the contact group on piracy. But the resolution has failed to address the root causes of piracy. On the contrary, pirates have started using sophisticated equipment which enables them to carry out attacks at further distances (up to 1500 nm) away from the coastlines. This has increased the costs of deterring piracy by the use of even more naval vessels and patrol boats. The effectiveness of such patrol boats too is questionable due to numerous policy and technical constraints placed on them by international organizations, such as the United Nations (UN), and due to the lack of sharing of actionable information between nations.

Furthermore, the various initiatives and efforts, such as rerouting of ships around the Cape of Good Hope, and some alternative solutions have been identified. The efficacy of deploying armed sea marshals on board vessels is a matter of intense debate that is prevalent in the shipping world currently.

Given the current trends, there is a likelihood that sophisticated submersibles may witness an emergence in the South Asian seas—to be used for the transportation of drugs— not too different from the situation in South America. However, the South Asian navies and Coast Guards are largely ill-equipped to counter such sophisticated technology and are hence not much of a deterrence to committed contraband carriers.

2.5 Maritime Pollution and Environmental Disasters

Pollution-related disasters at sea are a serious concern not only for the marine environmentalists, but for security analysts as well. While environmentally speaking they create a mayhem with the marine ecology, security concerns abound with the disaster affecting free flow of trade and shipping.

Countering transnational security threats requires cooperative approaches between the littorals and assistance of more capable countries in capacity building the other less capable maritime agencies. Maritime initiatives, such as the Indian Ocean Naval Symposium (IONS), provide the littorals with a forum for free discussion and cooperation along with similar platforms, such as IORA (Indian Ocean Rim Association), ASEAN Regional Forum (ARF), and the ADMM Plus. Other institutions that encourage cooperation are also active in this area. Finally, it is the political willingness to cooperate and help in terms of capacity building that really matters. These would eventually help in overcoming these transnational threats at sea and ensuring freedom of navigation, of which it is the fundamental pillar in supporting global and regional maritime security.

Chapter 3
Jurisdictional Issues of the Seas

Abstract The alarming aspect of Piracy has been on the increase at an alarming rate. In this chapter, it discusses the 'level of violence' of piracy, of which it is used to classify different types of piratical incidents, such as mugging, hijacking, barratry, and maritime frauds. This can then used to define crimes legally that can then be proceeded against in different courts of law.

Keywords Piracy · Violence · Somalia · International Navies

3.1 Modern Day Piracy

Not very much unlike the proverbial phoenix, the traditional form of piracy has manifested itself again in a new avatar at several locations of the world, particularly in East and West Africa. Incidents of piracy have been heard of in the Malacca Straits and South China Sea. Such incidents of piracy in the Eastern part of Africa at the mouth of the Red Sea have affected global maritime trade. This universal concern was eventually translated into a large naval presence along the Horn of Africa. Piracy can be categorized in different ways, by geographical classifications, by the intensity of attacks, or by the differing rationale or *raison d'être* for such attacks. The most accepted method depends on the geographical area where the attacks have happened.

The alarming aspect has been the increase in the level and intensity of violence in almost all incidents of piracy regional types of piracy without exception. Hence, it is the level of violence, which is used to classify different types of piratical incidents such as mugging, hijacking, barratry, and maritime frauds, that really matters. Indeed, an understanding of the level of violence can be used to define crimes legally and be proceeded against in different courts of law. In most piracy incidents, the captured crews are mistreated so as to pressurize ship/cargo owners to negotiate at the earliest as possible. The big bosses of the pirates also act as financiers and negotiators, apart from laundering the ransom amounts. They often fund the entire

© The Author(s), under exclusive license to Springer Nature Switzerland AG 2023   13
G. C. Gujar, A. K. Y. Ng, *Blue Economy and Smart Sea Transport Systems*,
SpringerBriefs in Geography, https://doi.org/10.1007/978-3-031-21634-3_3

logistical arrangement including acquisition of sophisticated weaponry, bribing the officials, and intelligence gathering.

The actual process of capturing a ship by armed pirates using several high-speed skiffs, together known as swarm tactics (all emanating from the nearby mother ship)—usually takes place at night. After a successful takeover of the ship, the pirates quickly steer the captive ship back to a safe harbor for the ransom demand and kickstarts the negotiation phase. Usually, the ransom amount runs into a few tens of million US dollars (depending on the age of ship and the value of the cargo she is carrying) with companies rarely revealing as to the precise sum being paid.

3.2 Organized Crime on High Seas

There are strong criminal linkages between globally organized crime syndicates and the pirates, according to Andrew Mwangura, who heads the East African Seafarers' Assistance Program. The extensive intelligence networks of the organized crime syndicates enable the pirates to have updated knowledge of the cargo and ships movements in the designated areas of their influence.

As most of the manufactured (high value) cargoes today are shipped in containers, it may not be farfetched to presume that pirates may choose to just intercept a few specific containers, instead of the entire ship. This does encounters some logistical problems involved in unloading containers or the cargoes at sea. Despite such, the scenario has considerable potential, as the returns are likely to be much higher than the more traditional means.

3.3 Patrolling the High Seas and Jurisdictional Issues

After considerable efforts in January 2009, the UN adopted Resolution 1851(2009) which called for the establishment of a contact group on piracy. However, the resolution was limited in resolving the problem as it largely failed to address the root causes of piracy that involved more complex socio-economic reasons, notably the poverty of communities that hosted the pirates. On the contrary, it sometimes opened the floodgates for naval/military actions—which despite some initial successes was ineffective in the long run.

In such circumstances, the usual tendency of most governments was to use the heavy hand of naval and military forces to resolve the problem, instead of countering it at the socio-political level. As such, the entire area became a melee of warships. The primary objective of such task forces was to deter, if not totally prevent, piracy attacks and ensure the safe passage of ships. However, often due to amorphous rules of engagement and unclear jurisdiction with regards to captured pirates, it becomes difficult to prosecute or imprison them. Furthermore, attempts to rely on military forces to solve piracy issue could attract bad publicity, particularly by different human rights groups around the world.

Another most preferred solution against piracy is to arm and man merchant ships by professionals called Sea Marshalls who also bear arms subject to the laws of the flag state. This is because of the fear of increased violence and secondary/collateral damage.

In addition, the US has declared piracy as a threat to the national security and forbade payment of ransom or rendering any kind of assistance to the pirates. Despite the strong recommendation from the shipping industry to universally legalize the deployment on board of Sea Marshall, the subject remains deeply controversial.

3.4 The Way Forward

It is obvious that the piracy problem is multidimensional and standard antipiracy responses, such as increasing naval presence, has not been entirely successful in eliminating the problem. The fact that this is more of a socio-economic and political issue rather than purely a security or military issue. As such, the policy development for tackling piracy anywhere can be viewed from the following perspectives:

- There is a critical necessity to incubate a strong government and develop and implement suitable welfare policies.
- The local populace is an aggrieved lot who have developed persecution complex vis-à-vis foreign governments in general and Western powers.
- Piracy demands the financial and logistical support links behind the actual pirates, which must be broken post haste.
- There is a burning need to strengthen legal support systems.
- Antipiracy patrols and the naval escorting of merchant ships in convoy must continue.
- A commonality in rules of engagement by warships in the area will ensure a unity in approach.

Chapter 4
Coastal and Port Security

Abstract Coastal security can be defined as protecting a country's coasts by securing the adjacent sea against the activities of state or nonstate actors/criminal groups. This apart, the problem was further compounded by policymakers all over who usually suffer from varying degrees of "sea blindness" issues regarding coastal security. Thus, it was often left to the customs, immigration officials, and coast guard to tackle the threats posed by smugglers and illegal immigrants. It was only after the terrorist threats on economic hubs and the humungous illegal immigration in several European nations that the world finally recognized the need for improving the coastal security system, and this chapter's focus on addressing this issue.

Keywords Coastal Security · National Jurisdiction · UNCLOS · Coast Guard

4.1 The Importance of Coastal Security

Any terrorist attacks on a country leaves a deep impact on the nation as it takes place in the most bizarre fashion in the economic centers of the country; may it be London, New York City, Paris, Brussels or Mumbai. These are all coastal cities and hence are more vulnerable to external attacks than other cities located in the interior. Such attacks expose the, in general, apathetic attitude of the governments and the security apparatus. The deep shortcomings in defending the coasts of the country were unveiled.

Coastal security can be understood as protecting a country's coasts by securing the adjacent sea against the activities of state or nonstate actors/criminal groups. This apart, the problem was further compounded by policymakers all over who usually suffer from varying degrees of "sea blindness" issues regarding coastal security. Furthermore, this issue has been considered as under the purview of the Navies or the Coast Guard. In addition, the focus, to the detriment of all else, was on securing the land borders, while the coastal regions attracted scarce attention from the defense policy makers. It was left to the customs, immigration officials, and coast guard to tackle the threats posed by smugglers and illegal immigrants. It was only after the terrorist onslaught on the economic centers, like Mumbai, and the humungous illegal immigration in several southern European nations the world woke up to the need for improving the coastal security system.

It is estimated that, on a single day, the coastal waters of most nations witness the passage of large number of floating crafts. Comparing such large numbers of vessels passing by, with the scarcity of resources (e.g., financial, technological, skilled manpower), it is not difficult to grasp the enormity of the problem faced by the governments all over.

Perhaps, the maximum damage can be inflicted by the nonstate actors/terrorists on the population centers along the coast and vital installations. The second concern is the damage caused by organized criminal gangs carrying out smuggling of narcotics, arms, and explosives. The existence of such criminal gangs is an important cause of serious domestic security concern because they also indulge in the illegal inflow of migrants.

4.2   Earlier Attempts at Coastal Security

The first maritime incident that caught global attention was the attack on USS Cole of the coast of Yemen followed by the bombing of US embassies in Kenya and Uganda and culminating in the 9/11 attacks on the World Trade Centre Towers in New York City. The response of the US Government was to increase extensive patrolling by warships all along the US coast and to launch the Coastal Security Scheme. The scheme envisaged the establishment of a series of coastal police stations along the coasts with the objective to strengthen patrolling along the entire coastal waters. However, most coastal state governments remained unenthusiastic due to financial constraints.

Another hurdle was more psychological in nature. This was the mindset of the local police constabulary, which insisted that their core job was to deal with land-based issues, and it was the duty of the US federal government to provide coastal security with the help of the coast guard and the national Navy. Due to lack of intelligence cooperation and coordination in addition to ineffective maritime border control and handling of the anti-terrorist operations, the respective national governments adopted a multipronged approach to tackle coastal security at various levels whereby the local marine police were granted jurisdiction up to 12 nautical miles from the shorelines, the coast guard was given the responsibility up to 200 nautical miles, and the national navies beyond 200 nautical miles.

4.3   Measures to Upgrade Coastal Security

The following consists of measures initiated by several of the governments to improve coastal security:

1. All coastal states were directed to (a) expedite the implementation of the approved Coastal Security Scheme, including early completion of construction of coastal police stations, check posts, outposts all along the coast, (b) immediately start using the services of local fishing boats/trawlers for patrolling and intelligence gathering, and (c) improve the coastal security in consultation with coast guard and local constabulary.
2. Improve and streamline the registration process of all types of all floating vessels.
3. To issue identity (ID) cards to all the fishermen.

4.4   Efforts by the National Navies

According to several US naval sources, the major problems facing coastal security include the lack of integration, coordination, and exchange of information between various agencies that involve various aspects of coastal security. This necessitates the setting up of 24×7 joint operation centers manned by specially trained personnel and networked with agencies like the Customs and Intelligence Bureau. The centers should be equipped with fast interception craft.

Apart from Naval and Coast Guard personnel, adequate and appropriate maritime training (or rather the lack of it) was found to be the weakest link for all other organizations; hence, it was decided to impart special training to the personnel manning these centers. The responsibility for training was handed over to the Navy. In addition, the lack of access to human and technical intelligence sources was identified as one of the weak links that has contributed to the various terrorist attacks. As such, measures were instituted to draft the huge fishing communities in coastal areas as the "eyes and ears" of the coastal security scheme. Additional budgets were allocated for this purpose.

The coastal radar chain that currently exists has been supplemented by the national automatic identification system (NAIS) network wherein a network of several electro-optic sensors is installed on lighthouses to track and monitor maritime vessels by receiving feeds from AIS transponders installed in these vessels. The system also facilitates the exchange of information between vessels as well as between a vessel and a shore station, thereby improving situational awareness and traffic management in congested lanes.

Apart from the sheer enormity of the coastlines and attendant issues, the task of patrolling such a large area is, to say the least, daunting. Also, it so happens that the federal and state governments perceive the coastal security policies from differing perspectives. Hence, its implementation at the state level often receives lower priority. This must be rectified by creating adequate awareness.

4.5   The Insecure Port

Port security can be understood as the security of all cargoes, passengers, users, and crew within a port. It can also include fighting crime within the port domain, such as smuggling, drug trafficking, counterterrorism activities, to name but a few. In a nutshell, it involves the protection of port facilities apart from the protection and coordination of security activities when ship and port interact. While not trying to downplay the contribution of previous port security-related studies, we posit that most existing research related to this topic are technical in nature (e.g., operational efficiency, IT security systems, calculating the cost of compliance with international guidelines). As such, comprehensive analysis on how such international guidelines can be applied in a regional or local perspective, including its problems, obstacles, and solutions, has remained scarce, particularly those which investigate port security in emerging economies. Given the importance of major ports in contemporary global economy, it was surprising that, after more than two decades since 9/11, such a gap has yet to be addressed satisfactorily. Hence, we focus on the policy changes that, according to us, are the immediate necessity.

4.6   Obstacles and Challenges

The lack of cooperation, the scarcity of financial and technical resources, and the embryonic stage of the risk management culture have ensured the stagnancy of security policies. As such, port authorities in almost all emerging economies are, remarkably, indifferent in complying with the International Ship and Port Facility Security (ISPS) Code.

One of the main reasons for such indifference in implementing security policies is that the standards stated therein are far too superior for local standards. This indicates that regional and local circumstances have been overlooked while formulating the security policy standards. Here it goes without saying that encouraging the development of the soft issues to avail themselves of local and regional cooperation, cross-institution information sharing, education, and know-how transfer and, most importantly, finance and other resources should receive higher priority. All these are pivotal in the creation of a robust coastal security system.

Most governmental organizations are loathe to share any actionable information with other appropriate organizations even on a need-to-know basis. This xenophobic attitude is cloaked under policy and red-tapism, but it also has got to preserving one's turf. The excuse of compromising security just simply does not hold water. In addition, many governmental organizations are often excessively centralized in decision-making related to security issues. Hence, any interactions with foreign organizations are normally done at a very senior level; the decisions taken by them just do not percolate to the lower ranks neither in words nor in spirits. Hence, we highlight the importance of having real-time exchange of information taking place at appropriate levels. A paradigm shift in attitude is needed to enhance coastal and port security in achieving the Smart sea transport system.

Chapter 5
Blue Economy and Smart Sea Transport Systems

Abstract This chapter discusses the Blue Economy, Smart sea transport systems, and the multifaceted challenges they are likely to face while attempting to develop such systems. Two novel ideas emerged in the latter half of twentieth century. The first idea of globalization was aggressively promoted, while the second idea of limits to growth was largely glossed over. It was only in 2008 that the second idea got its due in the face of the financial tsunami. Naturally, some changes had to take place. Technology was deployed across the board to identify and eliminate waste, improve efficiency, and reduce if not eliminate the effects of global warming and pollution. It was colloquially termed as Blue Economy. Simultaneously, the sea transport systems involved in global trade too had to be upgraded with similar objectives and thus the birth of Smart sea transport systems.

Keywords Global trade · Blue economy · Smart sea systems · Technology

5.1 The Growth of Global Trade and "The Box"

This chapter discusses the Blue Economy and the Smart sea transport systems and the multifaceted challenges they are likely to face while attempting to develop such systems. Two novel ideas emerged in the latter half of twentieth century. The first idea of "globalization" was aggressively promoted while the second idea of "limits to growth" was largely glossed over. It was only in 2008 that the second idea got its due during the financial tsunami in the first decade of this century. Naturally, some changes had to take place. Technology was deployed across the board to identify and eliminate waste, improve efficiency, and reduce if not eliminate the effects of global warming and pollution. It was colloquially termed as Blue Economy. Simultaneously, the sea transport systems involved in global trade too had to be upgraded with similar objectives and thus the birth of Smart sea transport systems.

Joseph Stiglitz, the Nobel Laurette in economics, suggests that both the stated forces of globalization, along with limits to growth, will perforce cause the mutation of sectors, regions, and countries to a higher evolved state and those entities, which do not do so, will shrivel and die, while others will survive and succeed. This evolved state of economy is termed the "Blue Economy." It focuses on various economic activities centered on the seas whereby the sea is perceived as a medium, which connects distant lands rather than separating them. Furthermore, economic activities based on the seas such as deep-sea mining, oil extraction, fishing, and tourism apart from transport will form the backbone of national economies. Such economies will emphasize lower costs, reduced waste, greater efficiency, and value addition in both manufacturing and services. Also, it implies less production, reduced consumption, and near or local sourcing. Most importantly, it signifies greater maritime security due to better manageability.

With almost 800 million twenty-foot equivalent units (TEUs) of containers being transported annually (UNCTAD, 2019/2021), the manufacturing and assembling operations have got further outsourced and have moved offshore, particularly to coastal regions which have been designated as special economic zones. Such global offshoring has contributed to the establishment and consolidation of today's global supply chains that largely shape the global production system and consumption patterns nowadays. In this regard, the special economic zones have provided numerous kinds of incentives and benefits to users. In recent times, several important issues connected with the manufacturing in special economic zones and subsequent shipping of manufactured goods and raw materials have drawn attention of national governments all over the world.

In this case, an important development took place, namely, the birth of megasized containerships which aggressively drove many decisions in the general direction of the 'trunk-and-feeder' shipping networks. This adds to costs and time delays rendering the small ports uncompetitive. The main driver of this phenomenon is the reduction in the average cost per container transported and handled, which is lower for larger vessels due to economies of scale in construction, manpower, and energy usage.

The outsourcing manufacturing activities has direct consequences on the type of merchandise shipped via maritime transport means. While maritime transport traditionally moved predominantly raw materials and agricultural products, it has now shifted to consumer and intermediate products. This trend heavily depends on improved process efficiencies, which, in turn, lead to the evolved state of the *Blue Economy*. Such an economy demands a secure maritime domain to thrive and succeed. This is especially true given the recent development of rival economic blocs (e.g., Regional Comprehensive Economic Partnership (RCEP), Indo-Pacific Economic Framework for Prosperity (IPEF)) and the evolvement from global supply chains to put more emphasis on 'global value chains'.

5.2 Changing Trends and the Birth of Smart Sea Transport Systems

It is impossible to ignore the impacts of global warming caused by rapid industrialization and excessive use of fossil fuels. Also, the side effects of the Covid-19 pandemic cannot be brushed under the carpet especially when the economies of most nations have gone for a toss. The ship owners in their quest to cut operating costs and offset the impact of higher fuel prices had no option but to further increase the size of ships to avail themselves of scale advantages. At the same time, port operators adopted the strategy of enhancing their port capacities by spending on dredging operations and improving last mile rail and road hinterland connectivity.

The ports have also modified the scope of their business models as they gradually shift their primary focus from shipping lines alone to the big shippers, consignees, and freight forwarders. To attract them, they have started (and more recently accelerated) providing value-added services, such as warehousing, inventory control, road, and rail services. In this case, the blockchain technology can provide additional solutions to address the issues of improving efficiency and providing the assurance, verifiability, and security of the information exchange that is critical in moving the global wheels of commerce efficiently and securely in the related areas.

5.3 Blockchain Application in Smart Shipping Operations

Another way to achieve and meet the blue economy and the Smart sea transport systems objectives is by using and leveraging the promise of new technologies. Hitherto, the maritime industry is still very traditional that heavily relies on physical documents and traditional methods to conduct daily business. Also, it is a highly regulated industry subject to IMO regulations, additional Flag State and Port State local regulations (e.g., Port State Control), the Classification Society rules, and other industrial standards, such as those set by Society of International Gas Tankers and Terminal Operators (SIGTTO), The Oil Companies International Marine Forum (OCIMF), and the International Safety Guide for Oil Tanker Terminals (ISGOTT). Ships are (re-)inspected almost endlessly by port and flag states, class, and industry inspectors such as OCIMF's Ship Inspection Report Program (SIRE) and Chemical Distribution Institute (CDI) inspectors.

This results in inefficiencies and increasing costs or a drop in the vessel safety and service quality. For example, the administrative time and effort spent by every ship's senior officer in storing, recording, and reproducing the same information, documents, and certificates that are demanded repeatedly by inspectors in every port. Ship owners and operators respond to this challenge by using digital systems and record keeping, but this has introduced other issues such as verifiability of the records and threats due to cyber security issues. In this regard, innovative technology, notably blockchains, can provide the solutions and the assurance, verifiability, and security of this information exchange that is critical in moving the global wheels of commerce efficiently and securely. A blockchain is an open distributed ledger that can record transactions between any two parties in an efficient, permanent, and verifiable manner. It is a growing list of records, termed as blocks, linked together by cryptographic hash of previous transaction and a time stamp. In our view, it can do this in three areas:

- It is conservatively estimated that on an average, approximately 30% of all working time on board is spent by the ship's senior officers on administrative tasks alone. This is precious time taken away from managing the safe and efficient operation of the ship and its maintenance. All this time spent by ship's senior officers to satisfy various inspectors in each port adds to the already burdened crew who is strapped for time and increases costs due to additional manpower requirements or, even worse, can result in crew fatigue that can cause incidents and accidents. The shipboard management of the vessel must maintain originals of several certificates, such as the International Oil Pollution Prevention Certificate (IOPP), the Continuous synopsis record (CSR), to name a few, and other accurate records such as the crew experience matrix, entries in the oil record book, up-to-date training records, inspection, and certification records of key equipment such as mooring ropes, lifting gear, and so on. A distributed, verifiable, and trusted network based on blockchain can be used to store such records and certificates. The Ship Inspection Report Programme (SIRE), Chemical Distribution Institute (CDI) inspectors, and port state inspectors can verify this information at any time from any location. The use of a blockchain-based system can enable real-time updates and a faster processing time of such records and documents. Automating this activity, which is currently performed manually, will not only improve document security, accuracy, verification, and transparency, but also reduce administrative costs and manpower requirements both ashore and on board and contribute to increased safety and quality without an associated cost increase. Imagine a SIRE/CDI inspection taking only four hours (instead of the current eight hours) in collecting, verifying, and recording information from the ship's records. The information stored on a blockchain would be visible to all interested market participants, hence enabling trust among them. The use of encrypted technology would enhance security from fraudulent activities, such as document manipulations, and allow for open, secure, and more efficient approach to the data management. Finally, it would reduce the involvement of any intermediaries allowing regulators and inspectors and other interested participants to directly communicate with the vessel and/or the owner/operator to gain access to the required data and records.
- Another novel area where blockchain can be used with huge benefits is in meeting the requirements of the recently implemented the Monitoring, Reporting, and Verification (MRV) requirements of the EU and the soon-to-follow IMO requirements on the vessel emissions. The MRV regulations require every major vessel to report its emission in a standardized way. This is the first step in the IMO's attempt to record and better understand global carbon dioxide (CO_2) emission from the shipping industry and use that in designing strategies and plans to help reduce this massive footprint. Using blockchain for this purpose can increase transparency, accuracy, verifiability, and reduce intermediary costs for regulators and shipping companies and others involved in this chain of activity.
- Global shipping in the twenty-first century is characterized by a move away from traditional, paper-based charts to electronic ones. Starting from July 2018, all vessels are required by the IMO to use the Electronic Chart Display and

Information System (ECDIS). The ECDIS uses the Global Position System (GPS) that shares and transfers data and software via the internet. Hence, it is vulnerable to cyber-attacks due to vulnerabilities in the system. As shipping moves toward cloud-based services and use and deployment of equipment and sensors that are connected to the internet (i.e., Internet of Things (IoT) devices), the risk of cyber-security and cyber-attacks increases. Blockchain-based distributed networks bear the advantage of countering cyber-security threats more effectively, compared to a centralized single database system. While malicious computer code and viruses can replicate and move across networks by exposing or exploiting the vulnerability in the system, the same technique will be very difficult, if not impossible, in blockchains that run on a distributed network.

5.4 Smart Port Development

Many ports in developing countries and emerging economies are characterized by obsolete equipment, hierarchical organizational structures, and an institutional framework that does not fit government's overall economic objectives. The governments too are not too keen to allocate scarce resources required for port development. They would rather hope for the private sector to chip in. But the private sector too is unenthusiastic because of rigid labor laws and regulations along with higher quanta of investments required. This enhances the risk to their capital tremendously. Hence the private sector desires port policy and other institutional reforms. However, given the slow pace of economic growth in the altered post covid-19 global economic circumstances, it is doubtful if the private sector would rise to the occasion, even if they possess the capacity to do so.

Since container traffic is expected to grow by about 4% in this decade (2020–2029), various governments aggressively try to pursue port and allied infrastructure development. Hence, the governments of developing countries, to facilitate foreign direct investment (FDI), in this sector have permitted 100% FDI and full autonomy. The ports sector managed to attract USD 66 billion between 2000 and 2014 (UNCTAD, 2019/2021), but it had all dried up since 2014 and has picked up at a brisk pace since the outbreak of the Covid-19 pandemic. This suggests that collaborations between public and private sectors are necessary to achieve the blue economy and the Smart sea transport system.

Chapter 6
Toward a Sustainable Global Growth

Abstract The old idea of collective security was amended, and the new concepts of common, comprehensive, and cooperative security were deemed as essential. This was bound to create confusion as security denotes and describes different things or conditions in different contexts depending on different perspectives. What are the implications of security failure and its implications for policy? Global security is more concerned with the systemic factors that influence the security and behaviors of states. In addition, it foresees the consequential implications regarding the interstate relations. This is because the state is regarded to be the primarily responsible actor for security in the international system. In this context, this chapter addresses the issues that are closely related to sustainable global growth.

Keywords Economic growth · International trade · Pollution · Environmental damage

6.1 The Current Security Architecture

In the late 1980s, the old idea of collective security was amended, and the new concepts of common, comprehensive, and cooperative security were added to essentially by the US and UK. This was bound to create confusion as security denotes and describes different things or conditions in different contexts depending on different perspectives. Most importantly, what are the implications of security failure and its implications for policy?

Global security is more concerned with the systemic factors that influence the security and behaviors of states. In addition, it tends to foresee the consequential implications regarding the interstate relations. This is because the state is regarded to be the primary responsible actor for security in the international system.

In the Indian Ocean Region (IOR), in general, it is obvious that the region is lively due to highly political interference by almost all global powers of substance, notably US, China apart from the Europeans, Indians, and the Japanese. The jostle for power is real whereby all the participating players are seeking to enhance their strategic influence.

At present, India and China are increasingly keen to undertake the leadership and responsibilities in global security, particularly in Asia. This responsibility was earlier discharged by the US and the UK before them. This necessarily means endeavoring to develop both strategy and a methodology to implement it.

With the US maritime power in erosion due to a lack of focus and change in priorities in the past decade, maritime disputes on the rise, the world cannot take the security of the maritime commons for granted. Neither is the world in opposition to resist or control the changes that are bound to take place. Hence, a rule-based order in the region becomes critical in the security architecture nowadays.

6.2   Altered Global Environment and Global Commons

It becomes evident that the regional security environment has become more important after the Sino-US relations have turned pear-shaped, which has led to the formation of several global political alliances and groupings, such as the Quadrilateral Security Dialogue (QUAD) and the Five Eyes (EVEY). The development of artificial islands in the South China Sea by China has further exacerbated the problem. The result is the imminent danger of national rivalries and resource competition threatening to encroach on the international freedom of navigation.

The main reason for this scenario is the global diffusion of maritime power because of the "rise of the rest" (particularly, India, Australia, and China), above all, which tends to alter the geostrategic maritime balance. This encourages all and sundry countries to project power beyond their territorial waters. In such circumstances, neighboring countries attempt to respond by strengthening their own naval forces, if they could afford or get into strategic alliances with the major powers.

In this case, the exploitation of marine resources indicates accessing and monitoring disputed and strategic waters. This results in littoral states and vested interests attempting to stake claims to vast areas adjacent to their coasts. This has added potential for further politicization and competition among the coastal countries. In addition, the nonstate actors, such as pirates, terrorists, and criminal syndicates possess the potential to complicate such matters further. Apart from the failure of the flag states with different policies to regulate this problem, security failure encourages the use of private maritime security companies (PMSCs). Consequently, the flow of trade is constrained and adversely affects the economic development of a state.

In the forthcoming years, maritime terrorism is likely to manifest and evolve itself in many unique ways that range from smuggling contraband to human trafficking to piracy and terrorist attacks in all its myriad manifestations. The use of the seas as a supply chain link for terror attacks on land-based targets is likely to be the ideal method of terrorist outfits, while the seas ensure the easy passage of men and material for terrorist attacks. Hence, the constabulary functions of maritime agencies, such as coast guards and marine police patrols, are likely to increase with the growing demands for a robust coastal security system.

6.3 Securing the Maritime Commons Through "Rules-Based Order"

The world is heavily dependent on the maritime ecosystem for trade and exploitation of maritime resources. As such, maritime interests can only be secured by assured access and sustainable management of the maritime domain. Moreover, the supply of energy and raw materials from the global marketplace is vital for the well-functioning of the global economy, as we can already witness from the energy crisis triggered by the recent Russo-Ukrainian conflict. This makes secured access to the maritime domain critical. As such, the maintenance of open and uninterrupted access of the stated sea lanes is vital for global trade and commerce. There are several narrow sea passages in Asia, such as the Gulf of Aden, Hormuz, and Malacca These potential chokepoints can easily be blocked by state and nonstate actors, either deliberate or by accident (e.g., the blocking of the Suez Canal by the containership Ever Given in March 2021). The mere potential for disruptions encourages local friction and global power competition.

6.4 Importance of Maritime Governance

It can be said that four distinct maritime governance scenarios can be discerned in the future. They are firstly an emphasis on global governance where multilateral frameworks and problem-solving mechanisms are made use of. This is followed by a 'maritime block scenario' where interstate conflicts with an intention to grab maximum power dominate. Thirdly, we may witness not only more frictions but also more effective regional governance, and lastly a contest for maritime dominance between state and nonstate actors.

India has since 2005 become a crucial ally of the US and has joined the QUAD. Also, it has been participating in the annual naval exercises in the Indian Ocean and elsewhere. The main objective is to restrain China from acquiring influence in the Indian Ocean Region, of which India considers it as its own backyard. In the bargain, India has become an important facet of the US "pivot or rebuilding strategy in Asia". The strategic partnership between the two countries is based on convergent geopolitical interests.

Countering the above-mentioned threats and challenges requires cooperation, understanding, and sensitivity of all stakeholders to security concerns of other countries. This requires a paradigm shift in maritime strategies and global vision of all the parties concerned. In this context, a matrix of cooperation should be developed that would enhance maritime bonding at different levels either singly, bilaterally, or multilaterally. This would not only help in overcoming the challenges and threats in the oceanic dimension, but also ensure the freedom of navigation for the efficient trade flows.

Hence, the next questions that demand further research include the identification of factors that make the blue economy possible and how to secure the gains of the resulting economic growth. A good example of the blue economy in action is the proposed "Greater Bay Area" in southern China that is planned to comprise of Hong Kong, Macau, Zhuhai, and other selected cities and regions in the Guangdong Province. Further investigation in this area can help us to envisage the role of the seas connecting these hub cities. A blue economy assists in converting the role of the sea from a separator to a connector and promotes the generation of wealth. Also, there is the question of securing wealth—who should play the role of the protector and at what costs? Who should bear the costs and in what capacity? Who owns the connecting sea and whose laws should prevail in the international waters? Finally, the question of illegal migration and its accompanying laws and policies too become relevant, especially when economic activities are generating substantial wealth. All these highlight that the necessity and, indeed, urgency of developing a generally-accepted naval doctrine that is critical to achieve sustainable global growth.

Chapter 7
Maritime Clusters

Abstract This chapter focuses on maritime clusters. Maritime cluster is an agglomeration of maritime-related industries concentrated in a specific region. Numerous marine-related industries, such as shipping companies, charterers, shipping financiers, Shipyards, marine insurers, arbitration bodies, maritime lawyers, port companies, inland transporters, warehouses, skilled labor, and environmental experts acting out of a single region that helps in achieving competitive advantages. It discusses the development of maritime clusters, notably their roles in industrial growth, connectivity, and international trade.

Keywords Maritime clusters · Industrial growth · Connectivity · International trade

7.1 Objectives and Success Factors

According to Porter (1998), "The global economic growth today is entirely a function of competitive advantage and the Ports act as Growth Poles and promote trade and Globalization." He was building on the Classical Evolution Theory propounded by Weber earlier and Theroux later that ports act as catalysts for regional economic growth. This birth and evolution of ports begins by agglomeration of core industrial activities in a specific region, which then grows to form industrial cluster formation. With further growth, the industrial clusters get converted into industrial hubs. The society rules and local government laws act as innovation enablers as well as trade promoters. Eventually, Trade Hubs are born (e.g., Hong Kong, Shanghai, Singapore, New York City). The next step is with the acquisition of scale and competitive advantage, the expansion extends into the regional coastal region.

It should be understood that industrial clusters are artificially constructed by identifying and encouraging growth of existing core economic specialist activities and are not born on their own. The core of such clusters comprises of economic specialist units, which are not dependent on any other activities. This core activity is concentrated in a geographically and spatially concentrated region. The strength

G. C. Gujar, A. K. Y. Ng, *Blue Economy and Smart Sea Transport Systems*,
SpringerBriefs in Geography, https://doi.org/10.1007/978-3-031-21634-3_7

of the backward and forward linkages with related organizations is critical for the survival of the cluster. Increase in physical distances weakens the linkages. Such clusters could be concentrated locally, regionally or beyond depending on the strength of the linkages. The clusters comprise of private business units, public-private entities like universities/research institutions, and purely public bodies like ports, airport authorities. The chief characteristics of such clusters are as follows:

1. Presence of skilled, large labor pool.
2. Possible to impart specialized training and upgradation in specific skills cheaply.
3. Presence of numerous suppliers and customers of specialized goods and services, thus reducing transport costs.
4. Enhanced interaction and autonomous self-governance results in creation of trust, reliability, and knowledge spillovers.

7.2   What Is a Maritime Cluster?

Maritime cluster is an agglomeration of maritime-related industries concentrated in a specific region. Examples of such clusters are the Port of Rotterdam region of the Netherlands and the Pearl River Delta region of China (or more recently, the 'Greater Bay Area' that comprised of Hong Kong, Macau, and part of the Guangdong Province). Numerous marine-related industries, such as shipping companies, charterers, Shipping financiers, Shipyards, marine insurers, arbitration bodies, maritime lawyers, port companies, inland transporters, warehouses, skilled labor, and environmental experts acting out of a single region that helps in achieving competitive advantages.

The phase-wise methodology adopted for development of a maritime cluster usually follows the undermentioned path:

1. It begins with enhancement of port capacities and productivity by adopting several standard reforms, such as formation of port authorities, handing over day-to-day management to an expert port operator on revenue sharing basis, dilution of government shareholding in the port, and eventually, selling off/leasing the port to the private operator.
2. It is usually accompanied by enhancement and improvement of port connectivity by deploying multimodal transport such as rail, road, pipelines, and waterways depending on the cargoes and geographical lay of the land. It is common that the port connectivity operations are conducted by private players who usually invest large amounts of money in developing the required infrastructure.
3. Nowadays, it is not unusual to find development of airports adjacent to ports. Such airports are usually linked to ports by dedicated and fenced road corridors to provide seamless connectivity for both air and marine cargoes.
4. Simultaneously, the development of port-dependent industries, such as steel plants, coal and oil power station, refineries, ship building, repair and ship breaking yards, food processing, fisheries, and other value-added industries is also

undertaken by the government itself or handed out to the private sector. Various concessions and subsidies are given to encourage such development.

5. It is equally important to upgrade the quality of the local labor in the adjacent coastal zones by skilling, educating, and recruiting people. It is obvious that this is not a short duration process, and a suitable long-term plan needs to be sketched out involving numerous steps such as conducting an initial survey of the local populace with a view to identify their strengths and weaknesses. It will be followed up by setting up purpose-specific schools, training institutes, and universities to teach specialized courses and evaluate the progress on continuous basis. Such students have to undergo practical internships/apprenticeships to acquire practical experience. Subsequently, they have to be integrated into suitable employment programs.

6. It would be important to encourage coastal tourism by creating some tourist attractions, like parks, resorts, or natural animal habitats to stimulate the interest of regional population and beyond to visit the said area and interact with local populace and generate some economic activities. A good example is the building of the tallest statue of Sardar Valla Bhai Patel (the first Home Minister of Independent India) in the world in a remote place in Gujarat and connecting it with rest of the country by road, rail, and air. Necessary infrastructure by way of hotels, restaurants, and airports were also being built.

Apart from imparting a boost to local industry, the development of clusters results in the innovation of various kinds, some of which are listed below:

1. Sustainable Water Transport.
2. Biotech applications to marine organisms.
3. Wind and Wave Energy power generation.
4. Fisheries and Sea Farming.
5. Artificial Meat products.
6. Water desalination.
7. Deep sea mining.
8. Carbon capture and fixing.

These innovations have been noticed in the Chinese Maritime Clusters in the Pearl River Delta, the Dutch maritime clusters in the Rotterdam area, as well as the Venice-Milan Italian Cluster and the Silicon Valley of the US.

7.3 Evolution of Maritime Clusters

It is noted that with greater change in cargo profile by way of value addition, scale economics, and sophistication, there will be a significant increase in the movement of such goods. Nowadays, with the birth and growth of trading platforms (e.g., Amazon, Alibaba, *Taobao*), it is not difficult to penetrate global markets. In addition, with the easy availability of revolutionized transport technology, which is bigger, faster, and more efficient, it is not difficult to get into a positive spiral of enhanced production scales and global trade. China has proved this point in various sectors. Such development of maritime clusters, coupled with a strong assistance from improved connectivity in information and infrastructure spheres, has brought in remarkable change in several countries in the Middle East and Latin America. It is obvious to one and all that the regions/cities that have adopted this process, such as Shanghai, Hong Kong, and Singapore have benefitted and evolved, while those that refused to adapt to this demand of change (e.g., Kolkata, Cape Town, Manchester) have lost value.

Under the Belt and Road Initiatives (BRI), China has been involved in setting up maritime clusters in several regions of the world. Some of the Chinese BRI projects are listed hereunder. Apart from providing a major boost to regional economies, it also increases the influence of China in these maritime clusters:

1. Eurasian Land Bridge: Bypasses Suez Canal and Reduces transit time by half.
2. Gwadar–Karakoram–Sinkiang–Western China: Bypasses Malacca Straits.
3. Kyaukphyu, Myanmar–Kunming, Southern China: Increases Competitive Advantage of Chinese Goods.
4. Hambantota, Sri Lanka: Development of Major Hub Port on Europe–Far East Trade would force transshipment for Indian cargoes.
5. Hong Kong–Guangzhou High Speed Rail: Integration of seven ports and nine airports in the "Greater Bay Area".

The major objectives of developing maritime clusters could be manifold, some of which are listed below:

1. Inter-Industry Cooperation leads to innovation and enhancement of product quality.
2. Improves Customer Satisfaction and enhances demand.
3. Optimizes Inventory, thus lowering inventory carrying costs.
4. Improved Forecasting leads to better planning, improved efficiency, and elimination of waste.
5. Employees Satisfaction leads to increased productivity.
6. Achieving competitive advantage results in greater trade, revenue, and profits.

The anatomy of maritime cluster has different body parts, like information collectors, market researchers, manufacturers, sales, distribution, etc., which respond to different demands like markets, finance, environment, politics by cooperating and collaboration in different measures.

7.4 Cluster Synergy

There are various types of interactions taking place within and beyond the cluster simultaneously. Some are minor and takes place regularly, while some are major, which take place less frequently. However, there is little doubt that such interactions result in an exchange of value and enhancement of trust. Trust is the key ingredient for sustained growth. Some of the synergy-creating transactions are listed below:

1. Low interaction, Low collaboration—department-specific functions, e.g., Purchase of tires is responsibility of Transport department.
2. High interaction, Low collaboration—Higher levels of documentation & procedures are necessary, for example, specific product lines—lipsticks, mascara, eye-liners in a cosmetic firm.
3. Low interaction, High Collaboration—where markets are rapidly changing for example, consumer electronics.
4. High interaction, High Collaboration—critical customer-specific products—NASA, ESA, Boeing.
5. Imbalance between the two concepts viz., Collaboration and Interaction will result in loss of focus and affixing of responsibility.

7.5 Organizational Clusters

The clusters are usually organized by connecting different kinds of mini-organizations created for different purposes and activities together by secured linkages. A brief idea of such a cluster is given here as under:

1. Consists of different organizations linked together.
2. Linkages are used to transmit goods, information, knowledge, services, and personnel.
3. Important for Cluster Formation (e.g., Silicon Valley, Toyota City).
4. Benefits of corporate & local level co-ordination.
5. Results in attaining global production scales and faster technology changes.

7.6 Synergy Creation and Organization Management

Synergy creation is one of the ultimate goals of cluster formation. It is essentially a byproduct of trustful interactions. It requires fast, efficient, effective conflict and dispute resolution mechanisms. Some of the steps necessary for such a phenomenon are as given under:

1. Synergy in this context essentially means pooling of different thought processes.
2. This is achieved by way of healthy dialogue.
3. This also means recognition and respecting differences of opinion.
4. Have a route map for consensus and common vision.
5. Reconcile the differences.
6. Integrate different thought processes.

7.7   Inferences from Successful Cluster Formation

It is natural to expect that such maritime clusters will be highly automated, easily scalable, and of course will consume as little energy as possible. Thus, it is reasonable to expect the clusters to be equipped with artificial intelligence (AI) to exploit the new opportunities rising on the horizon. This will enable fast decision-making process and equally fast error correction mechanisms. Such 'Smart' clusters are expected to be deployed with the following equipment, hardware, and software:

1. Installing Trillions of sensors to collect information.
2. Superfast Computer Analysis and Disseminating information in real time to relevant stakeholders.
3. Developing IT connectivity (e.g., data collection, storage, transfer).
4. Focus on Continuous Training Program.
5. Greater Automation, Artificial, and Machine Intelligence.

As stated above, the most critical ingredients for creation of successful maritime clusters are transparency and trust. This will naturally be facilitated by the governments/state. In addition, the government will have to invest massively to create necessary infrastructure. The government is expected to focus on the following aspects:

1. Infrastructure development.
2. Land acquisition.
3. Waterways and dredging.
4. Port, Airport construction.
5. Cheap water and electricity.
6. Cheap capital.
7. Reasonable labor laws.
8. Roads and Railways.
9. Universities, Trade and Skill Training.
10. Tax moratorium.
11. Duty-free import of capital goods.
12. Maritime Security.
13. Foreign collaborations.
14. Export Guarantees.

Indeed, it is obvious that not all the governments have the resources to invest in such risky projects. On the other hand, the private sector does not only have the resources but also the necessary knowledge and competency to execute such projects. However, it is beyond their ability to undertake such risky projects without the active support of the government. As such, there is a genuine need of real public private partnerships (PPPs). The PPPs will involve undertaking the following policy and legal measures:

1. Autonomous Cluster governance and reduced red tapes.
2. Low entry and exit barriers.

3. Agglomeration of heterogeneous industries.
4. Enhanced internal competition.
5. Easy availability of venture capital.
6. Flexible labor laws.
7. Rapid conflict resolution.
8. Acceptance of valid business failures—No black marks.

The spread of the Covid-19 pandemic and subsequent lockdowns and unequal pace of vaccination has led to remarkable rise in economic opportunities as well as in all other human spheres, such as health, education, and transport, to name but a few. This excessive rise in inequalities of all kinds has prompted the governments to open their purse strings and distribute critical necessities such as food, water, and medicine for free. Naturally, there is a limit to what the governments can do. As such, there is a need for the private sector to step in and pick their share of the responsibility to help humans survive. The measures that need to be undertaken can be found below.

1. *Greater Emphasis on*

   (a) Corporate Social responsibility.
   (b) Imparting skills and education to staff for greater productivity.
   (c) Water Conservation.
   (d) Sustainable power and transport.
   (e) Health, Lifestyle, and stress management.
   (f) Affordable housing.
   (g) Job Security.
   (h) Sharing Knowledge.

2. *Toward a Blue Economy*

   (a) Sea as a connector and not as a separator.
   (b) Exploiting marine resources in a sustainable manner.
   (c) Sustainability and green shipping.
   (d) Coastal development.
   (e) Strategic alliances with other maritime clusters.
   (f) Maritime Security.

It is obvious that various measures and policy changes need to be undertaken on war footing to reduce the inequality in the society all over, kick-start the engine of economy, prevent further spread of the pandemic, and focus on reducing carbon emissions and global warming. Some of the issues raised earlier in the chapter are reiterated below:

1. Complex Organizational and Managerial skills are necessary.
2. Compatibility potential creation for regional cluster tie-ups.
3. Leads to enhanced national competitive advantage.

4. Rise in Innovation, value addition, productivity, and economic growth.
5. Birth of revolutionary products.
6. Increased employment across all grades.
7. Improved living standards.
8. Creation of reliability and trust.

Chapter 8
Public Private Partnerships and Port-Land Interface Projects

Abstract This chapter illustrates the importance of port-land interface projects for the development of Blue Economy. It analyzes the current scenario—How and why of port-land interface policy and the future scenario—How and why it should be altered under the context of public-private partnerships. It ends with useful recommendations for further improvements.

Keywords Connectivity · Port-land interface projects · Monopoly · Public private partnerships

8.1 Current Scenario

The primary policy objective while developing such policy in port-land interface projects is to:

- To make a country's international trade cost competitive.
- To ascertain and affix the responsibility and liability for loss of customs revenue of the inland transporter while transporting EXIM cargoes to and from the gateway ports.
- To deploy a cargo compatible mode of transport.
- To ensure prevention of leakage, pilferage, damage, and loss of customs revenue.
- To expedite decongestion of the port in a cost-effective manner.
- To make the ports achieve competitive advantage by generating push and pull forces.

The secondary policy objective is to ensure the integration of the ports—Forward, Backward, and Sideways. This will facilitate the seamless integration of the various port users and stakeholders, such as:

G. C. Gujar, A. K. Y. Ng, *Blue Economy and Smart Sea Transport Systems*,
SpringerBriefs in Geography, https://doi.org/10.1007/978-3-031-21634-3_8

- Ocean carriers integrate into ports, inland terminals, and landside transport links.
- Multimodal operators integrate into the reverse of this chain.
- The railways, which combine with port terminals.
- Road operators, which become logistics service providers.
- It enables the freight forwarders to extend traditional service boundaries.

This is essentially a result of unimaginable growth of technology and Knowledge Economy. However, we still have a long way to go. In the last decade, there has been tremendous growth in the role of Mechanization and Automation in containerization in the ports and inland depots and freight stations. Some of the notable examples are as follows:

- Electrically operated Automated Guided Vehicles (AGV).
- Unmanned straddle carriers.
- AI assisted container yard planning and equipment optimization.
- AI assisted labor utilization for stuffing/destuffing.
- AI assisted hinterland transport capacity optimization and forecasting.
- Optical scanners and recorders for gate activity.
- Smart reefer charging stations.
- Automated warehouse operations and delivery.

Also, we witness usage of Automation/Mechanization in Bulk Cargoes Warehousing, Logistics, and Distribution. For instance, we can witness such a paradigm shift in power stations (coal), steel plants (iron ore), fertilizers (MOP/DAP), as follows:

- Modern, safe, dust free, and weather-proof silos/warehouses.
- Automated truck and wagon loading.
- Automatic bagging plants.
- Mechanized internal shifting by earth movers.
- Automatic sampling, testing, grading, and certification.
- Loss, damage, and theft mitigation using optical scanners and recorders.
- Automated and computerized gate operations.
- Computerized weigh bridges.
- Ultra-modern firefighting system.

Similarly, we witness automation/mechanization in handling of liquid bulk cargoes warehousing, logistics, and distribution in various refineries, tank farms, and food industries. Some notable examples are as follows:

- Centrally controlled, multiple products tank farms of various sizes and specifications.
- AI assisted forecasting and multiple utilization of tanks.
- Automated heating control.
- Mechanized cleaning of tanks, pipelines, and sludge disposal.
- Automated truck and wagon loading.
- Product delivery over long distances by using pipelines safely.
- Optical sensors used to prevent leakage, theft, contamination, and evaporation losses.

- Computerized gate operations.
- Automated sampling, quantity, reconciliation, and quality checks.

On the other hand, we can see the deployment of automation and mechanization in handling, storage, and distribution of perishables and hazardous cargoes, such as:

- Grading, sorting, debulking, cleaning, packing, labeling, certification, palletization, and handling of products.
- Refrigeration and quality control.
- Truck and Container stuffing and loading.
- Isolated storage and safe handling.
- Organizing auctions.
- Inventory control.
- Optical scanner controlled computerized gate operations.
  Some notable examples are handling and distribution of Halal products, cheese markets, flower markets, and paper waste.

Similarly, the deployment of automation and mechanization can be used for break-bulk cargo storage and distribution. Some examples are as follows:

- Overdimension and heavy lift cargoes.
- Heavy duty cranes.
- Trained operators and drivers.
- Special insurance.
- Special roads for inland transportation.
- Lashing/chocking/packaging.
- Special vehicles for inland transport.

Due to improved port connectivity, we currently witness the following changes:

- About 10% of total container throughput originates or is destined to Inland Container Depots (ICDs)/Container Freight Stations (CFSs).
- Most Container Traffic is imbalanced and concentrated (85%) in few hub ports serviced by large container carriers of 20,000 TEU capacity and above.
- About 60% of the cargoes are dispersed within 200 kms of the port's vicinity.
- Pipeline—30%—only for liquid cargoes.
- Rail—(25–30%)—Rail economical over longer distances, provided availability of infrastructure and competitive freight.
- Road—(70–75%)—Flexible, cost competitive, unorganized sector, pollution, congestion.
- Environmental impact—Not accounted for adequately.
- Last Mile Connectivity—A reason for bottlenecks.

Despite the growth in automation/mechanization, there are several obstacles and challenges, some of which are listed below:

- Government regulations.
- Poor rail, road connectivity.
- Lack of Supply Chain Management Expertise.

- Imbalance use of port capacity (especially JNPT).
- Lack of real-time logistics data tracking.
- Lack of cargo security.
- Insufficient and unsuitable warehousing facilities.
- Unreliable local transport.
- Poor web-based operations capability.
- Undisciplined and inefficient labours.

In addition, there is the issue of increased competition and costs. Among the developed countries, trade logistics costs are typically 10% of GDP. For developing countries, these costs often exceed 30%. For example, one kg of tea costs fewer than two USD at the production site in Assam and is sold at 72 USD per kg in retail. In developing countries, freight costs alone (transport and insurance) can make up to 40% of value of exports for several landlocked countries. As such, the total logistics costs (e.g., packaging, storage, transport, inventories, administration, management) are estimated to reach up to 20%.

The obvious solution is to achieve scale advantages by increasing capacities. But then, one needs to consider the quantum of risk and the dangers it entails. In most developing countries, a few ports have exhausted their capacities, while most have huge underutilization of capacities. These developments involve capital-intensive technologies. However, due to the risk of formation of monopolies/oligopolies, there is a need for adequate antitrust legislation and competition regime. But it should be remembered that scale requires monopoly creation and monopoly demands an appropriate pricing strategy. In addition, larger ships impose huge investments in deeper and bigger ports, piers, and faster evacuation. Thus, ports need to work in tandem with requisite landside facilities and connectivity entail highly capital-intensive infrastructure.

To realize scale advantages, an inland thrust needs to be generated by deploying push and pull forces. This will assist in realizing the hinterland potential. Container traffic is estimated to be at least 70%; however, the actual movement of full containers from and to hinterland locations currently is often less than 50%. Container movement between dry ports and gateway ports currently is in 32% range, an optimal ratio is at least 50%. For the demand of adequate port infrastructures, overwhelming volumes traverse other regional ports, like Colombo, Singapore, Dubai/Salalah—with resultant additional cost and transit time. This requires setting up of numerous dry ports all over countries and regions, especially those with huge landmass. Apart from the factors mentioned above, there are some other advantages of dry ports listed below:

- Quicker evacuation of containers from the ports.
- Quicker turnaround time for ships.
- Lesser congestion at ports.
- Speedy clearance and regulatory compliance saving time.
- Inventory management & redistribution on the just-in-time (JIT) basis.
- Advantage of Scale.
- Repacking and distribution.

- Manufacture/assembling.
- Utility of space.
- Utility of time.

Moreover, proper policies, which would regulate the competition among the various dry ports, need to be put in place. To do so, the government needs to attract multiple players by providing, for instance, tax incentives. Furthermore, it should ensure the non-formation of cartels and ensure that price/tariff ceilings are strictly adhered to, and ensure service quality and price flooring. More policy measures, such as ensuring regional allocation for different groups of participants, should be implemented by establishing an adequate number of dry ports set up in a balanced manner in all regions (so as to avoid or minimize overcrowding).

Furthermore, the average dwell time at ports and dry ports is 4.3 and 3.7 days, respectively. The reasons for this inordinately high dwell time are as follows:

I. Waiting for cargo consolidation.
II. Waiting for onward transport (e.g., ship, train, trucks).
III. Waiting for customs inspection and permissions.
IV. Waiting for labor to conduct stuffing/destuffing operations.
V. Waiting for empty containers.
VI. Waiting for equipment.
VII. Warehousing for duty payment based on market conditions.

In most developed countries, a seamless connection from ship to aircraft—The Air Freight Station (AFS) Concept—has been developed. The AFS concept is based on the idea of providing airlines' nodes near to the production centers. Activities traditionally performed at airports will be done at the AFS:

- Customs documentation and examination.
- Cargo acceptance check (weight/volume).
- Security checks (x-ray screening/physical check).
- Palletization.
- This will avoid congestion at the gateway airports.

8.2 Future Scenario

The ports of the future, if properly developed, will be a different beast altogether. They will become:

- Highly integrated with various stakeholders.
- Smart Ports- Requires sophisticated information technology and a vigorous, continuous training program.
- Greater automation—labor issues.
- Extremely capital intensive.
- Portal databases.
- Low transaction costs.

However, to develop such ports and address appropriate connectivity issues, the government requires the assistance of the private sector especially their managerial and marketing skills. It should be noted that privatization is not the panacea for addressing all the issues. This is so because of the following factors:

- Risk—Too Uncertain—Not Quantifiable.
- Returns—Too Low—Not Assured.
- Trust Deficit—Too Wide.
- Pricing Policies—Not Sustainable.
- Vision/Objectives—Different.
- Requirement of Capital—Too Large.
- Altered Global Economic Environment.

Hence, there is an urgent need for developing Public Private Partnerships (PPP). The reasons for doing so are as follows:

- Huge investments are required to create the necessary infrastructure.
- Inadequate public funding leading to significant gap in supply and demand.
- Need for attracting private capital and providing reasonable rates of return.
- Need for enhanced managerial efficiency.
- All stakeholders to share risks and rewards.
- Government should facilitate development of PPPs.
- Real-time data collection.
- Big data analysis.
- Policy development, implementation, and audit.
- Market regulation in a transparent manner.
- Development of organizational and managerial skills.
- Education and vision development - integration with global education and research networks.
- Mass communication of shared vision.

There is a strong need to develop PPPs in warehousing, distribution, and logistics. However, it entails the undertaking of the following measures by the governments:

- Land acquisition allocation, development, and leasing out.
- Providing sovereign guarantees, cheap capital, or equity participation.
- Joint marketing efforts.
- Assuring minimum quantum of state-owned cargoes.
- Provision of adequate insurance at reasonable premiums.
- Providing state security for cargoes, personnel, and cyber.
- Setting up of customs facility.
- Providing rail and road connectivity.
- Dredging of shallow stretches (in instances of river or seaside terminals).

In conclusion, one can state as under:

- Growth in asset values has far exceeded performance growth.
- New capacities coming online are much more expensive, hence would add to logistics costs making Indian products less competitive.
- Political influences will further distort markets, thereby hurting the end consumer.
- Complex organizational and managerial skills should be developed immediately for a paradigm shift.

Chapter 9
The Pandemic, the Blue Economy, and the Way Forward

Abstract The raging pandemic has adversely impacted the entire world to smaller or larger extent and has destroyed people and livelihoods too. It has put on holds plans for expansion and growth. Sustainable growth is almost nowhere on the mental horizons of policy makers all over. Pure survival of themselves and their brethren is all that matters to each one of us. Yet nobody can afford to ignore the looming impact of climate change, nor the need to feed the ever-growing populace neither the need to find some productive livelihood for the teeming millions. This is a tall order indeed when the governments all over do not know how to balance the conflicting needs and keep the said challenges at bay. This chapter discusses this dilemma.

Keywords Pandemic · Vaccine logistics · Maritime trade · Volatility

9.1 The Volatile Maritime Trade

The raging pandemic has adversely impacted the entire world to smaller or larger extent and has destroyed people and livelihoods too. It has put on holds plans for expansion and growth. Sustainable growth is almost nowhere on the mental horizons of policymakers all over. Pure survival of themselves and their brethren is all that matters. There are numerous countries tethering on the brink of bankruptcy, default, and chaos. Volatility is the order of the day, with prices of products and services experiencing extreme swings. Searching for certainty and reliability in these chaotic times appears to be running a fool's errand. Yet, nobody can afford to ignore the looming impact of climate change nor the need to feed the ever-growing populace neither the need to find some productive livelihood for the teeming millions. This necessitates the free movement of capital, goods, information, and labor. This is a tall order indeed when the governments all over do not know how to balance the conflicting needs and keep the Covid-19 virus at bay.

The consequences of the pandemic are here to stay with us for the foreseeable future. Novel methods for studying, working, and entertainment have been adopted and will probably be in vogue in some form or other for the foreseeable future. One of the consequences of the Covid-19 pandemic was the slowdown in global production and international trade. However, the collapse in global demand for manufactured goods did slow down the economic cycle. But it was of short-term nature as the demand soon picked up in the second half of 2020.

All said and done, the collapse of demand for certain kind of manufactured goods was compensated by the increasing needs of other commodities, particularly computer peripherals and allied equipment, mobile phones, and, of course, pharmaceutical goods and medical equipment. Furthermore, the consumption of inventory in the first half of 2020 during the various forced lockdowns was restocked in the second half of 2020.

The International Monetary Fund (IMF) forecasted that the world economy would grow by about 2%, in view of the emergence of developing countries and emerging economies, notably China. However, the Baltic and International Maritime Council (BIMCO) has warned that such higher trade growth would unlikely be sustainable as the effect of the pandemic fades away. All these predictions have largely come true. This was evidenced by the steep increase in freight rates and the historic profits made by the shipping industry. In this regard, container shipping industry did unexpectedly well by making handsome profits after a long spell of draught. Anecdotal evidence suggests that shippers around the world may need to pay freight rates that are several times more than usual. Furthermore, many are forced to queue up in ports for weeks, export or import their goods from Asia. In this regard, the shipbuilding industry in Mainland China, South Korea, Japan, and Taiwan too has hit a purple patch with their new order book bulging with a total of 147 containerships been ordered since October 2020, compared with just 40 ships in 2019.

The shipping lines initially reacted to the pandemic by "culling" of capacity from major trade lanes, colloquially termed as blank sailings. Numerous port calls were cancelled. It also led to lower service frequency and poor service quality. With the increase in call sizes, the laid-up tonnage rose as well. The issue of reduced supply was further aggravated by the steps adopted by lines (e.g., slower speed, longer routes). These allowed shipping lines to sustain freight rates at higher levels, forcing other stakeholders (e.g., shippers, transport associations) to express their discontent to the competition authorities in, say, the EU, US, China, and Australia.

Although regulatory bodies (e.g., the Federal Maritime Commission (FMC) of the US) started to look at the rapacious behavior of the container carrier industry, it is unlikely that they would take stern measures to curb this behavior. The reason for such leniency shown to the container carriers is the recognition by the regulators that excessive competition in certain industries would push prices down to marginal costs that would, eventually, lead to bankruptcies, further concentration, and monopolies. This is the chief reason for the conditional exemption granted to shipping alliances from antitrust laws. Such forms of cartel formation existed earlier too under the garb of shipping conferences. The only difference being the objective of the alliances is to generate profits via cost control, while liner shipping conferences strived for a similar objective via price-fixing. Whether the same principles apply to other transport industries, such as airlines, remain to be seen.

Hence, we can safely argue that shipping line cartels (or termed as conferences or alliances) are here to stay, at least in the foreseeable future. One may witness more mergers and acquisition activities in allied industries, especially those that involve vertical integration, such as warehousing, dry ports, inland road and rail transport, software development, and of course, port development, and governance. In fact, vertical integration has accelerated at an unprecedented rate since 2020 and seems to continue growing from strength to strength in the foreseeable future.

9.2 Impact on Ports

The throughputs of many major ports suffered in early 2020. However, the spectacular revival of demand in the second half of 2020 immediately led to a spike in demand for port services. This was mainly due to large-scale restocking, taking place in countries of the west in North America and Europe in the latter half of 2020. Indeed, such demands for port services have yet to peter out (UNCTAD, 2019/2021) and have even spread to different regions (e.g., from the west to the east coast ports in the USA). The ports were taken aback by this sudden spike in demand and were caught unprepared. The resilience of global supply chains was brought into serious question, having suffered from the scarcity in equipment, truckers, and dock labors. Another factor that demands attention is the severe shortage of shipping containers. This phenomenon too can be explained by the sudden precipitous fall in demands for shipping services in the first half of 2020, followed by the equally sudden spike in demands for the same services in the second half of the same year. The system could not adjust to the new demands, with containers left in the wrong places with no cargoes available to export and little time available to the carriers for repositioning empty containers. This happened chiefly due to the fact that many containers having been used in the first quarter of 2020 to move medical equipment to different parts of the world, such as Africa and Latin America. Simultaneously, the very high demands and freight payable by Asian shippers forced the carriers to reposition the empties at breakneck speed, to Asia without providing western exporters the capacity they desired to export goods to other places around the world.

The increase in average ship sizes has imposed a heavy burden on ports. The intention was to compensate for blank sailings and reduce frequencies (Cullinane & Haralambides, 2021). In this case, it can be said that the time to handle a container arriving on a larger ship is slightly longer as it imposes additional costs on the port by way of demanding additional equipment and more time. However, berth efficiency mainly depends on the availability of cranes that reach row 24–27 and beyond, i.e., in the increase of boom sizes and the distance the crane trolley must travel. The other things demanding the attention of the port operators simultaneously (e.g., gate congestion, dwell times, establishment of dry ports, modernization of customs services, re-handling, port equipment, allocation of berths) result in the bar being set too high for the port operator. Inevitably, the port performance declines on one parameter or other.

9.3 The Green Regulatory Regime

Before the Covid-19 pandemic in 2020, the shipping industry largely focused on the possible impacts of the IMO 2020 global Sulphur cap regulation, notably costs. The efficacy of scrubbers and very low Sulphur fuel oil (VLSFO) at higher cost was the topic on which the shipping companies focused. It inevitably led to stockpiling and higher prices. Subsequently, the prices of VLSFO collapsed due to the advent of Covid-19 and slump in demands for shipping services. Though VLSFO prices have recovered now to their original levels, the regulators have lost the ardor to chase the lines and force them to reduce emissions. In current circumstances, carbon emissions and pollution control appear to have been put on back burner; for how long one does not know.

9.4 What Does the Future Hold?

All the stated factors in conjunction with the spiking of freight rates have practically broken the backs of the trade in emerging economies, thus creating huge uncertainties and even despair. One just found it difficult on how to respond to the fast-changing situation, and then all one is left to do is fend of the emerging problem and stoically await the occurrence of the next. It is noted that the Covid-19 pandemic and its consequences have come as an external shock. Also, considering the efficacy of the vaccination program, the effects of the pandemic will most probably be temporary. As mentioned earlier, the forecasts are generally positive.

A particular economic issue is the astronomical amount earmarked in fighting against Covid-19 globally, especially for mitigating its effects on employment (Cullinane & Haralambides, 2021). Most of the governments around the world have issued unlimited debts to feed and clothe their populace while the economies have continued to collapse. This dual pincer has practically bankrupted a few countries and regions, and the only solution appears to be in wholesale debt write-offs by the western countries and the global financial institutions and issuing fresh debts. Naturally, this would involve political trade-offs and rising inflation and its collateral effects. Whether the world has the collective wisdom to handle this situation remains to be seen, especially with the rise of new conflicts (e.g., the Russo-Ukrainian War and its accompanied energy crisis) and the establishment of rival economic partnerships (e.g., RCEP, IPEF).

Finally, the effects of explosive growth in state-of-the-art technology (e.g., 3D printing, teleworking) taking place all over well-established concepts and "common sense" (e.g., localization, near-shoring) have found fertile ground. There is a clear intention of the current US (Joe Biden's) administration for the bringing back the industrial production home and minimizing the risk of overdependence on China, which would also benefit the US labour market. Such trade policies, if implemented, would pose negative impacts on transport, from air to maritime transport. The second aspect, which is difficult to ignore, is the ardent desire of US and Europe to wean itself off the overdependence on China (and Russian natural gas). Sooner rather than later the process will begin in the earnest, and the establishment of the blue economy and the Smart Sea Transport System are likely to be pivotal in battening down the hatches and smoothening the process.

Bibliography

Bichou, K. (2004). The ISPS code and the cost of port compliance: An initial logistics and supply chain framework for port security assessment and management. *Maritime Economics & Logistics, 6*(4), 322–348.

Cullinane, K., & Haralambides, H. (2021). Global trends in maritime and port economics: The COVID-19 pandemic and beyond: Editorial. *Maritime Economics and Logistics, 23*(3), 369–380.

Dobbs, R., & Manyika, J. (2015). *No ordinary disruption: The four global forces breaking all the trends* (pp. 132–133). Public Affairs Books.

ESRIF. (2009). *Working group: Security of critical infrastructures, final report, Part 2*. ESRIF.

European Commission (EC). (2009). *Report assessing the implementation of the directive on enhancing port security*. EC.

Flynn, S. E. (2006). Port security is still a house of cards. *Far East Economic Review*, Jan/Feb 2006.

Ghosh P. K. (2011). Indian Ocean dynamics: An Indian perspective. Accessible at: www.eastasiaforum.org

Ghosh, P. K. (2012). Somalian piracy: An alternate perspective, observer research forum, Occasional Paper #16.

Greenberg, M. D., Chalk, P., Willis, H. H., Khiko, I., & Ortiz, D. S. (2006). *Maritime terrorism: Risk and liability*. RAND.

Hanssen, B. (2009). Presentation on implications of global security. *ISPS Code on Dryports*. Dryport EU project.

Hariharan, K. V. (2001). *Containerization and multimodal transport in India*. Shroff Publications.

Hidekazu, I. (2002). Efficiency changes at major container ports in Japan: A window application of data envelopment analysis. *Review of Urban & Regional Development Studies, 14*(2), 133–152.

IMO. (2002). *Amendments to the annex to the international convention for the safety of life at sea, 1974 as amended*. IMO, SOLAS/CONF.5/32, adopted in December 2002.

IMO website.: http://www.imo.org. Last accessed on February 2008.

King, J. (2005). The security of merchant shipping. *Marine Policy, 29*, 235–245.

Kommerskallegium – The National Board of Trade, Sweden, *Supply chain security initiatives: A trade facilitation perspective*, 2008.

Levinson, M. (2006). *The box: How the shipping container made the world smaller and the world economy bigger*. Princeton University Press.

Limao, N., & Veneables, A. J. (1999). *Infrastructure, geographical disadvantage and transport costs*. World Bank.

© The Author(s), under exclusive license to Springer Nature Switzerland AG 2023
G. C. Gujar, A. K. Y. Ng, *Blue Economy and Smart Sea Transport Systems*,
SpringerBriefs in Geography, https://doi.org/10.1007/978-3-031-21634-3

Mahan, A. T. (1914). The Panama Canal and the distribution of the Fleet. *North American Review, 200.*

Mukherjee, P. K., & Basu, A. B. (2009). *Legal and economic analysis of service contracts under Rotterdam Rules.* World Maritime University, Malmo, Sweden, working papers.

Ng, A. K. Y. (2007). Port security and the competitiveness of short sea shipping in Europe: Implications and challenges. In K. Bichou, M. G. H. Bell, & A. Evans (Eds.), *Risk Management in Port Operations, logistics and supply chain security* (pp. 347–366). Informa.

Ng, A. K. Y., & Liu, J. J. (2014). *Port-focal logistics and global supply chains.* Palgrave Macmillan.

OECD. (2003). *Security in maritime transport: Risk factors and economic impact.* OECD Maritime Transport Committee.

Paul, J. (2005). India and the global container ports. *Maritime Economics and Logistics, 7*(2), 189–192.

Peterson, J., & Treat, A., *The Post-9/11 Global Framework for Cargo Security.*

Pilling, D., & Mitchell, T. (2006). US official urges Asia to improve port security. *Financial Times,* 28 March.

Porter, M. E. (1998). The Competitive advantage of nations. Palgrave Macmillan, London.

Reinhart, C. M., & Rogoff, K. S. (2010). *Growth in the time of Debt, American Economic Review,* papers and proceedings *573–578.*

Schoonen Advies en Management. (2009). *De Containerbinnenvaart als Secure Lane, Eindrapport.* Amsterdam.

Srikanth, S. N., & Venkataraman, R. (2007). Strategic risk management in ports. In K. Bichou, M. G. H. Bell, & A. Evans (Eds.), *Risk management in port operations, logistics and supply chain security* (pp. 335–345). Informa.

Stephens, B. (2015). *America in retreat: The new isolationism and the coming global disorder.* Sentinel.

Stopford, M. (2000). *Defining the future of shipping markets.* Presentation on Global Shipping Trends. ITIC Forum.

Tan, A. T. H. (2005). Singapore's approach to homeland security. In K. W. Chin & D. Singh (Eds.), *Southeast Asian affairs 2005* (pp. 329–362). ISAS.

Thornton, P., Ocasio, W., & Lounsbury, M. (2012). *The institutional logics perspective: A new approach to culture, structure and process.* Oxford Academic.

UNCTAD. (2019/2021). *Review of maritime transport.* UNCTAD.

UNESCAP. (2009). Transport and communications bulletin on Asia and Pacific No 78, pp. 102–110.

World Ocean View. (2010). A look at the future. Accessible at: http://worldoceanreview.com/en/wor-1/transport/

World Security Index website: http://worldsecurity-index.com. Last accessed in February 2020.

Woxenius, J., Roso, V., & Lumsden, K. (2004, September 22–26). The dry port concept – Connecting seaports with their hinterland by rail. *Proceedings of the First International Conference on Logistics Strategy for Ports,* Dalian, China.

Index